Troubleshooting Manual
LD 465-1 / LD 465-1C / LT 465-1C
LDS-465-1A / LDS 465-2
Engines
TM 9-2815-210-34-1
for
2.5 Ton 6x6 M44A2 Truck Series
aka
Deuce and a Half

April 1981

edited by
Brian Greul

The M44A2 series of military trucks is affectionately known as the deuce and a half, or simply the deuce. These ubiquitous trucks were first produced during WWII and General Eisenhower credited them as being one of the most important US Army vehicles. This book is the engine troublshooting manual for the LD-465 series of diesel engines. It is being printed for enthusiasts, restores, and collectors who may wish to own a quality paperback copy. The editor has endeavored to minimize changes, but the following changes were made: Feedback forms are omitted, the fonts have been updated for printing purposes on modern equipment.

Should you have suggestions or feedback on ways to improve this book please send email to Books@OcotilloPress.com

---

Edited 2021 Ocotillo Press
ISBN 978-1-954285-46-0

Printed in the United States of America

Ocotillo Press
Houston, TX 77017
Books@OcotilloPress.com

TM 9-2815-210-34-1
T.O. 38G1-48-12-1

TECHNICAL MANUAL

VOLUME 1 OF 2

TROUBLESHOOTING

DIRECT SUPPORT AND GENERAL SUPPORT LEVEL

ENGINE ASSEMBLY, DIESEL (MULTIFUEL):
NATURALLY ASPIRATED OR TURBOCHARGED,
FUEL-INJECTED, WATER-COOLED, 6-CYLINDER

MODELS: LD-465-1, NSN 2815-00-239-5824;
LD-465-1C, NSN 2815-00-134-4830; LDT-465-1C,
NSN 2815-00-103-2642; LDS-465-1, NSN 2815-00-075-0087;
LDS-465-1A, NSN 2815-00-239-5819; AND LDS-465-2,
NSN 2815-00-808-8011 AND CLUTCHES

NOTE:
THE STYLE OF THIS TM IS EXPERIMENTAL. IT IS BEING TRIED BY THE ARMY ONLY ON A LIMITED BASIS

Chapter 1 General Information
Chapter 2 Troubleshooting Approach
Chapter 3 Troubleshooting Index
Chapter 4 Test Equipment Procedures Index
Chapter 5 Troubleshooting Roadmap
Chapter 6 Fault Symptom Index
Chapter 7 Sample Troubleshooting Procedure
Chapter 8 Engine System Troubleshooting
Chapter 9 Engine System Support Diagrams
Chapter 10 Engine System Test Procedure

DEPARTMENTS OF THE ARMY AND THE AIR FORCE
APRIL 1981

WARNING

EXHAUST GASES CAN BE DEADLY

Exposure to exhaust gases produces symptoms of headache, dizziness, loss of muscular control, apparent drowsiness, and coma. Permanent brain damage or death can result from severe exposure.

Carbon monoxide occurs in the exhaust fumes of internal combustion engines, and becomes dangerously concentrated under conditions of inadequate ventilation. The following precautions must be observed to insure the safety of personnel whenever the engine is operated for maintenance purposes or tactical use.

Do not operate engine in an enclosed area unless it is adequately ventilated.

Do not idle engine for long periods without maintaining adequate ventilation.

Be alert at all times during engine operation for exhaust odors and exposure symptoms. If either are present, immediately ventilate working area. If symptoms persist, remove affected personnel from area and treat as follows: expose to fresh air; keep warm; do not permit physical exercise; if necessary, administer artificial respiration.

If exposed, seek prompt medical attention for possible delayed onset of acute lung congestion. Administer oxygen if available.

The best defense against exhaust gas poisoning is adequate ventilation.

WARNING

SERIOUS OR FATAL INJURY TO PERSONNEL MAY RESULT IF THE FOLLOWING INSTRUCTIONS ARE NOT COMPLIED WITH.

Diesel fuel is very flammable. Care must be used when choosing a place to work on fuel system. Keep engine about 50 feet away from an area where open flame, sparks, or smoking may cause a fire. Keep a fire extinguisher close by.

Fuel coming out of an injector nozzle under pressure can go through the skin. This can cause blood poisoning. Keep hands away from injector nozzle when doing test procedures.

Make sure engine has had at least an hour to cool down before working on cooling system. Due to high temperature of coolant, bad burns can occur if contact is made with skin.

* TM 9-2815-210-34-1
T.O.38G1-48-12-1

TECHNICAL MANUAL
NO. TM 9-2815-210-34-1
TECHNICAL ORDER
NO. 38G1-48-12-1

DEPARTMENTS OF THE ARMY
AND
THE AIR FORCE
WASHINGTON, DC, 3 April 1981

TECHNICAL MANUAL

VOLUME 1 OF 2

TROUBLESHOOTING

DIRECT SUPPORT AND GENERAL SUPPORT LEVEL

ENGINE ASSEMBLY, DIESEL (MULTIFUEL): NATURALLY ASPIRATED OR TURBOCHARGED, FUEL-INJECTED, WATER-COOLED, 6-CYLINDER

MODELS: LD-465-1, NSN 2815-00-239-5824;
LD-465-1C, NSN 2815-00-134-4830; LDT-465-1C,
NSN 2815-00-103-2642; LDS-465-1, NSN 2815-00-075-0087
LDS-465-1A, NSN 2815-00-239-5819; AND LDS-465-2,
NSN 2815-00-808-8011 AND CLUTCHES

REPORTING OF ERRORS AND RECOMMENDING IMPROVEMENTS

You can help improve this manual. If you find any mistakes or if you know of a way to improve the procedure, please let us know. Mail your letter, DA Form 2028 (Recommended Changes to Publications and Blank Forms),or DA Form 2028-2 located in the back of this manual direct to: Commander, US Army Tank-Automotive Command, ATTN: DRSTA-MB, Warren, Michigan 48090. A reply will be furnished to you.

TABLE OF CONTENTS

VOLUME 1 OF 2

	Paragraph	Page
CHAPTER 1. GENERAL INFORMATION		
Scope	1-1	1-1
Organization	1-2	1-1

*This manual together with TM 9-2815-210-34-2-1, 3 April 1981; and TM 9-2815-210-34-2-2, 3 April 1981, supersedes TM 9-2815-210-34, 16 June 1978.

TABLE OF CONTENTS-CONT

		Paragraph	Page
	Troubleshooting Approach	1-3	1-1
CHAPTER 2.	TROUBLESHOOTING APPROACH		
	General Approach	2-1	2-1
	Troubleshooting Index	2-2	2-1
	Test Equipment Procedures Index	2-3	2-1
	Troubleshooting Roadmap	2-4	2-1
	Fault Symptom Index	2-5	2-1
	Sample Troubleshooting Procedure	2-6	2-1
CHAPTER 3.	TROUBLESHOOTING INDEX		
	General	3-1	3-1
	Index	3-2	3-1
CHAPTER 4.	TEST EQUIPMENT PROCEDURES INDEX		
	General	4-1	4-1
	Index	4-2	4-1
CHAPTER 5.	TROUBLESHOOTING ROADMAP		
	General	5-1	5-1
	Roadmap	5-2	5-1
CHAPTER 6.	FAULT SYMPTOM INDEXES		
	General	6-1	6-1
	Index	6-2	6-1
CHAPTER 7.	SAMPLE TROUBLESHOOTING PROCEDURE		
	General	7-1	7-1
	Sample Detailed Procedure	7-2	7-1
CHAPTER 8.	ENGINE SYSTEM TROUBLESHOOTING		
	Equipment Items Covered	8-1	8-1
	Equipment Items Not Covered	8-2	8-1
CHAPTER 9.	ENGINE SYSTEM SUPPORT DIAGRAMS		
	General	9-1	9-1
CHAPTER 10.	ENGINE SYSTEM TEST PROCEDURE		
	General	10-1	10-1
	Test Set-up	10-2	10-1
	Test Procedure	10-3	10-1

VOLUME 2 OF 2

Part 1 of 2

(TM 9-2815-210-34-2-1)

		Page
CHAPTER 1.	GENERAL MAINTENANCE INFORMATION	
CHAPTER 2.	PRELIMINARY PROCEDURES FOR ENGINE MAINTENANCE	
Section I.	Scope	2-1
Section II.	Preparation of Engine for Repair	2-1

TABLE OF CONTENTS-CONT

		Paragraph	Page
CHAPTER 3.	ENGINE DISASSEMBLY		
Section I.	Scope		3-1
Section II.	Removal of Engine Accessories		3-1
Section III.	Disassembly of Engine Assembly		3-129

Part 2 Of 2

(TM 9-2815-210-34-2-2)

		Paragraph	Page
CHAPTER 4.	CLEANING, INSPECTION, AND REPAIR		
Section I.	Scope		4-1
Section II.	General Cleaning, Inspection, and Repair Procedures		4-1
Section III.	Cleaning, Inspection, and Repair of Engine Components and Accessories		4-2
CHAPTER 5.	FINAL ASSEMBLY		
Section I.	Scope		5-1
Section II.	Assembly of Engine Assembly		5-1
CHAPTER 6.	DYNAMOMETER TESTS AND RUN-IN PROCEDURES		
Section I.	Scope		6-1
Section II.	Run-In, Tests, and Adjustments		6-1
APPENDIX A.	REFERENCES		A-1
INDEX			Index 1

CHAPTER 1

GENERAL INFORMATION

1-1. SCOPE. This volume shows you how to do troubleshooting at the direct support and general support levels of maintenance. The amount of troubleshooting you can do is based on what the Maintenance Allocation Chart says you can fix. Because of this, the only trouble symptoms you will find here are those that could be caused by faulty things you can fix.

1-2. ORGANIZATION. This volume has the information you will need to troubleshoot the engine. Chapter 2 tells you how to use the information in the other chapters of this volume to find what is wrong with the truck, and what you must do to fix it. Chapter 7 has a procedure that takes you step-by-step through a sample troubleshooting procedure and shows you how to use the information to find the trouble and fix it.

1-3. TROUBLESHOOTING APPROACH. In order to find out what is causing the problem in the truck, you must use a good approach. A good approach just means a way of doing troubleshooting so you can find the problem and not get confused or lost. The following chapter describes how you can use the materials in this volume to troubleshoot with a good approach.

CHAPTER 2

TROUBLESHOOTING APPROACH

2-1. GENERAL APPROACH. This chapter gives you instructions on how to use the troubleshooting material to help, you find and fix the trouble. In every system of the engine there can be faults or problems which will cause certain symptoms. Symptoms can be such things as unusual noise, vibration, or even complete failure of a system. This volume gives information for each system on which you can do troubleshooting to find faults and fix them. Before you troubleshoot a system, you should look at the troubleshooting indexes which will lead you to the information you need to help make your troubleshooting faster and easier. If you follow the instructions the right way, you will find those troubles you can fix. But, if you fix something and the trouble is still there, it means there is more than one trouble. If this happens, start all over again to find the other trouble.

2-2. TROUBLESHOOTING INDEX. The troubleshooting index, and instructions on how to use it are in chapter 3. Go to this index first because it tells you where to find troubleshooting roadmaps, fault symptom indexes, summary troubleshooting charts and support diagrams for each system.

2-3. TEST EQUIPMENT PROCEDURES INDEX. The test equipment procedures index, and instructions on how to use it are in chapter 4. This index tells you where to find tests which you can use to do your troubleshooting. It also tells you what equipment you will need to do the tests. If you have a STE/ICE (Simplified Test Equipment/ Internal Combustion Engine) Set (NSN 4910-00-124-2554), you may use it, where applicable, to do your troubleshooting. Refer to TM 9-4910-571-34&P.

2-4. TROUBLESHOOTING ROADMAP. A troubleshooting roadmap is given in chapter 5. If the system is made up of subsystems, these subsystems are also on the roadmap. Under the subsystem is a list of things which are the most likely causes of a fault symptom in that subsystem. If you have enough skill, you can troubleshoot these things on the engine without using the detailed troubleshooting procedures. So if you know enough about the engine to work on your own, use the roadmap for the system with the problem before you check the fault symptom index.

2-5. FAULT SYMPTOM INDEX. Fault symptom indexes and instructions on how to use them are in chapter 6. For each system of the engine, there is an index which gives you a list of the fault symptoms for that system. The index also tells you where to find the detailed troubleshooting procedures and what resources (tools/ people) you need to do each procedure.

2-6. SAMPLE TROUBLESHOOTING PROCEDURE. A sample troubleshooting procedure is in chapter 7. This sample procedure will help you see the way detailed troubleshooting procedures are to be used.

CHAPTER 3

TROUBLESHOOTING INDEX

3-1. GENERAL. This chapter has a troubleshooting index which covers every system of the engine on which you can do troubleshooting. The index tells you where to find all the other information you need to do your troubleshooting procedures.

3-2. INDEX. The troubleshooting index (figure 3-1) is divided into five columns that list the system, troubleshooting roadmap, fault symptoms, summary troubleshooting procedures, and system support diagrams. The following breakdown tells you what is in each column.

a. System Column. This column gives a list of systems on the engine for which troubleshooting can be done at the direct support and general support maintenance levels.

b. Troubleshooting Roadmaps Column. This column tells you where to find the troubleshooting roadmap for each listed system. These roadmaps are given in chapter 5.

c. Fault Symptom Index Column. This column tells you where to find the troubleshooting fault symptom index for each listed system. Fault symptom indexes are given in chapter 6.

d. Summary Troubleshooting Procedures Column. This column tells you where to find the summary troubleshooting procedure for each listed system. Some systems do not have summary troubleshooting procedures, so the column will be left blank for those systems.

e. System Support Diagrams Column. This column tells you where to find support diagrams for each listed system. Some systems do not have support diagrams, so the column will be left blank for those systems.

	SYSTEM	TROUBLE-SHOOTING ROADMAP	FAULT SYMPTOM INDEX	SUMMARY TROUBLE-SHOOTING PROCEDURES	SYSTEM SUPPORT DIAGRAMS
1	ENGINE	Figure 5-1	Table 6-1		Figure 9-1-9-4

TA 120830

Figure 3-1. Troubleshooting Index

CHAPTER 4

TEST EQUIPMENT PROCEDURES INDEX

4-1. GENERAL. This chapter has a test equipment procedures index which tells you where to find the tests you need to do your troubleshooting.

4-2. INDEX. The test equipment procedures index is divided into three columns that list test equipment, tests, and figure numbers. The following breakdown tells you what is in each column.

a. Test Equipment Column. This column tells you what kind of equipment you need to do your troubleshooting tests.

b. Tests Column. This column tells you what tests are given in this manual. Next to each piece of test equipment are listed the tests that you can do with that equipment. This column also gives troubleshooting tests which can be done without using test equipment.

c. Figure Column. This column tells you where you can find the tests in this manual.

TEST EQUIPMENT		TESTS	FIGURE
1	____________	Fuel Injector Nozzle	10-1
2			
3			
4			
5			
6			
7			
8			

TA 120831

Figure 4-1. Test Equipment Procedures Index

CHAPTER 5

TROUBLESHOOTING ROADMAP

5-1. GENERAL. This chapter gives a troubleshooting roadmap for the engine for which you have detailed troubleshooting procedures. Figure 5-1 covers the roadmap for the detailed procedures.

5-2. ROADMAP . The roadmap gives a list of things which are most likely to cause a fault symptom in a system or subsystem. At least one of the items listed will be found to be bad when you do the detailed troubleshooting procedures for that system.

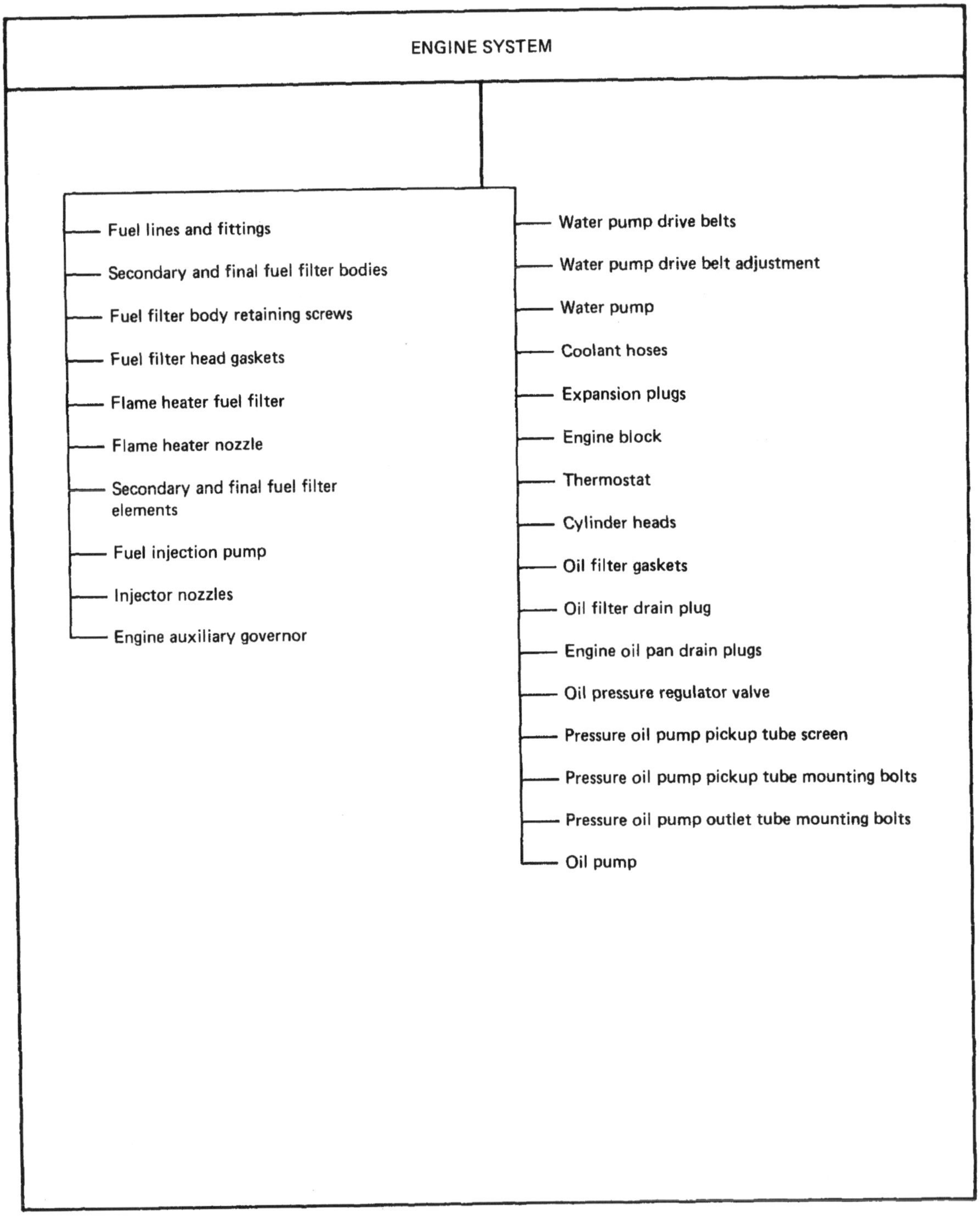

TA 120832

Figure 5-1. Troubleshooting Roadmap, Engine System

CHAPTER 6

FAULT SYMPTOM INDEX

6-1. GENERAL. This chapter gives a troubleshooting fault symptom index for every system of the engine for which you have detailed troubleshooting procedures. The index is in table form (table 6-1) which gives you a quick way to check what material you have to use to do your troubleshooting.

6-2. INDEX. The index is divided into columns which give you information you need to help you do troubleshooting procedures. The following breakdown tells you what is in each column.

a. Subsystem Column. If the main system is divided into subsystems, the subsystems will be listed in this column.

b. Symptom Column. This column lists the symptoms, or problems for which detailed troubleshooting procedures are given.

c. Summary Column. This column tells you where to find the summary troubleshooting procedures for each symptom.

d. Detailed Column. This column tells you where to find the detailed troubleshooting procedure for each symptom.

e. Persons Column. This column tells you how many people are needed to do the troubleshooting procedure.

f. Special Tools Column. Any tools needed to do the troubleshooting procedure which are not included in your common tool kit are listed in this column.

g. Standard Tools Column. A dot in this column means that tools found in your common tool kit are needed to do the troubleshooting procedure.

h. Materials Column. This column tells you what materials are needed to do the troubleshooting procedure. These materials and how they will be issued will be decide by your maintenance officer.

i. Time Column. This column tells you how much time you will need to do the detailed troubleshooting procedure. The time will be decided by your maintenance officer.

FAULT SYMPTOM INDEX

TABLE 6-1. ENGINE SYSTEM

SUBSYSTEM	SYMPTOM	TS PROCEDURE		RESOURCES REQ'D				
		SUMMARY	DETAILED	PERSONS	TEST EQUIPMENT		MATERIALS	TIME
					SPECIAL TOOLS	STANDARD TOOLS		
——	1. Engine cranks but does not start	——	Figure 8-1	2	——	•		
——	2. Engine runs too fast	——	Figure 8-2	1	——	•		
——	3. Water temperature gage reads above 200°F while running	——	Figure 8-3	1	——	•		
——	4. Low oil pressure	——	Figure 8-4	1	——	•		
——	5. Engine lacks power	——	Figure 8-5	1	——	•		

TA 120833

CHAPTER 7

SAMPLE TROUBLESHOOTING PROCEDURE

7-1. GENERAL. This chapter gives a sample troubleshooting procedure. The purpose of the sample procedure is to help you see how detailed troubleshooting procedures, test equipment procedures, and summary troubleshooting procedures are used to find faults in a system.

7-2. SAMPLE DETAILED PROCEDURE. (See figure 7-1.) The sample detailed procedure given is the fuel system troubleshooting procedure for the symptom, STARTER MOTOR WILL NOT CRANK ENGINE. This symptom is one you will have when you try to start your truck and certain parts on the truck are not working correctly. In each numbered box, instructions are given which tell you what to do, and how to do it. A large dot is placed next to the "what-to-do" instructions, and small dots next to the "how-to-do-it" instructions.

a. Box number (1) gives general instructions on getting the engine ready before you start to troubleshoot.

b. Box number (2) gives fault isolation test instructions. In this case you are told to check starting system circuit for loose, burned, or broken leads and connections. These tests or checks are often referred to in detailed troubleshooting procedures to help you find the problem and fix it. After you do the tests or checks you read the question at the bottom of box number (2). If the starter system is not okay, the answer to the question is (NO), so you go to the next box.

c. Box number (3) gives you a corrective action. In this case the fault is burned or broken leads or connectors. The corrective action is what you do to fix the fault, which is to replace any burned or broken leads or connectors. If the engine still doesn't start after you do this, it could mean that there are other faults in the system. When this happens, go back to the beginning of the procedure and do each step again until you find the other faults.

d. Sometimes the corrective actions given for a fault will tell you what to do to fix the fault, but will not give you detailed instructions on how to fix it. Instead, you will be told to refer to another volume in this manual for these instructions. Box number (4) is an example of this.

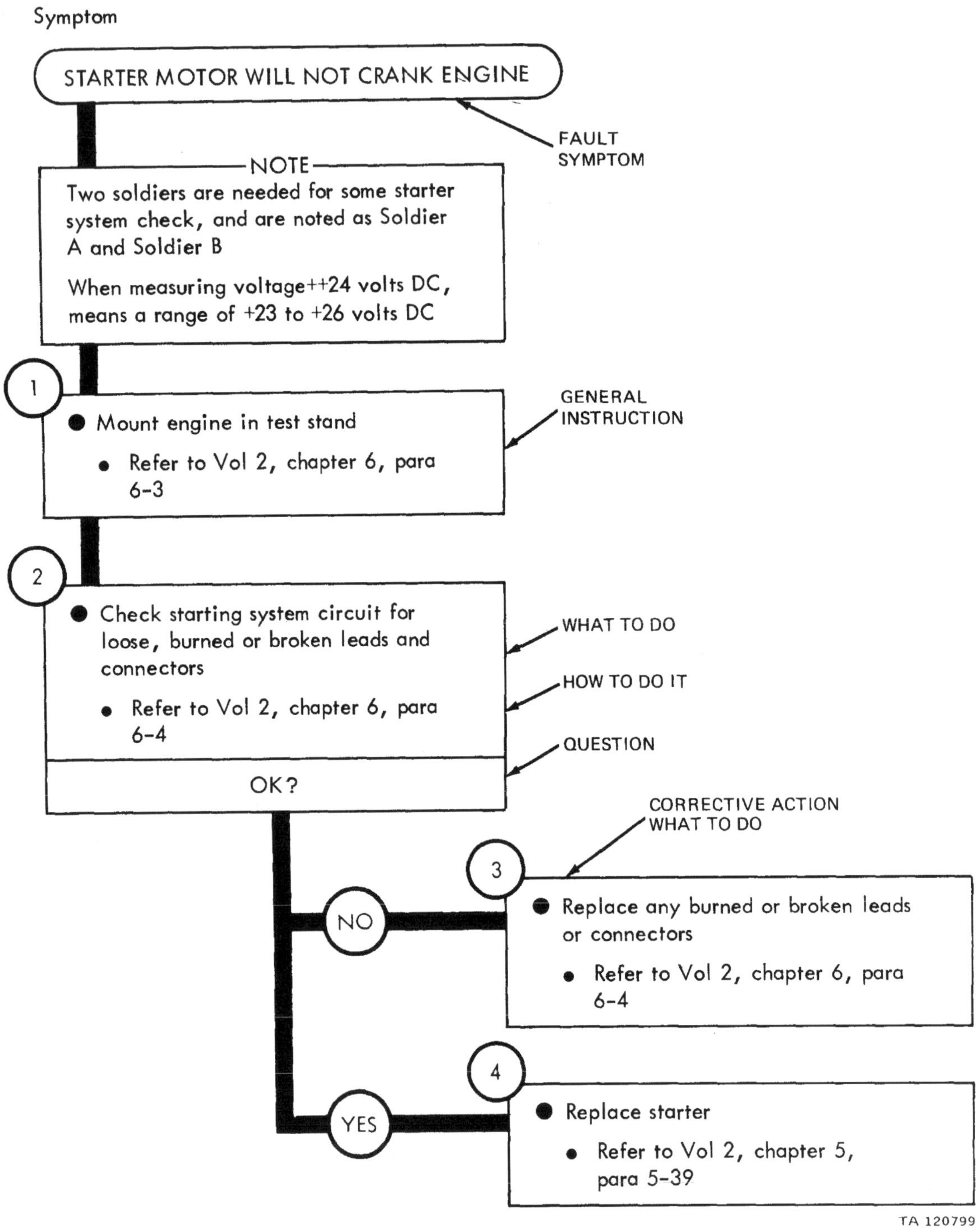

Figure 7-1. Sample Troubleshooting Procedure

CHAPTER 8

ENGINE SYSTEM TROUBLESHOOTING

8-1. EQUIPMENT ITEMS COVERED. This chapter gives equipment troubleshooting procedures for the engine system, for which there are authorized corrective maintenance tasks at the direct support and general support maintenance levels.

8-2. EQUIPMENT ITEMS NOT COVERED. All equipment items for which corrective maintenance is authorized at the direct support and general support maintenance levels are covered in this chapter.

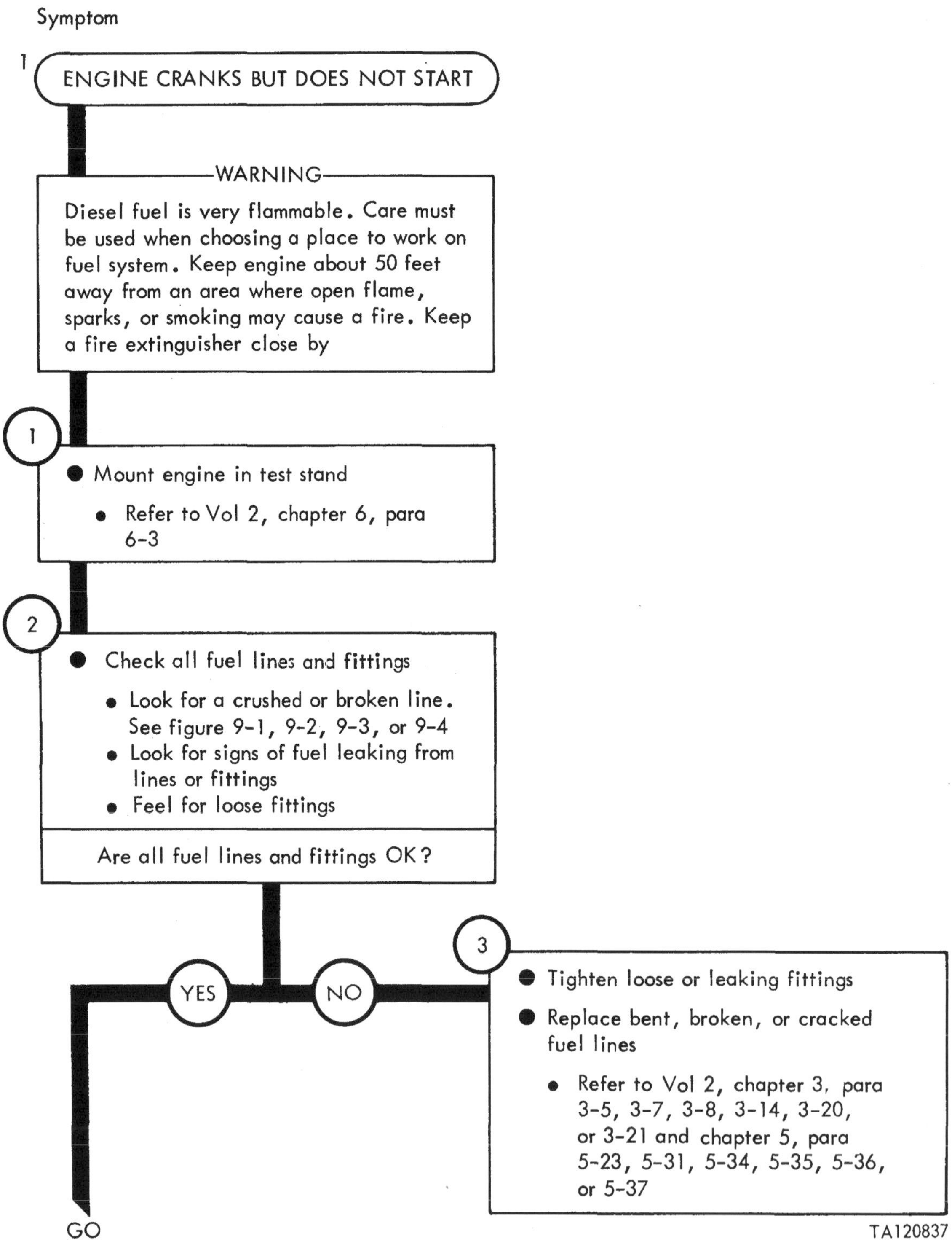

Fgure 8-1 (sheet 1 of 7)

GO

4

- Check secondary and final fuel filter bodies
  - Look for fuel leaking from fuel filter bodies
  - Look for signs of fuel leaking from fuel filter head gaskets
  - Feel for loose fuel filter retaining screws

Are fuel filter bodies OK?

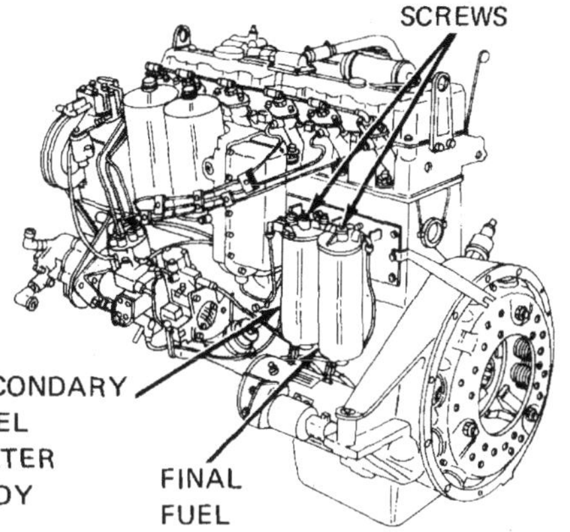

YES NO

5

- Tighten loose fuel filter retaining screws
- Replace leaking fuel filter head gaskets
  - Refer to Vol 2, chapter 3, para 3-14 and chapter 4, para 4-33 and chapter 5, para 5-31
- Replace leaking fuel filter bodies
  - Refer to Vol 2, chapter 3, para 3-14, chapter 4, para 4-33, and chapter 5, para 5-31

6

- Check flame heater fuel filter
  - Remove flame heater fuel filter. Refer to Vol 2, chapter 3, para 3-21, chapter 4, para 4-72 or 4-79
  - Look for signs of dirt in filter

Is flame heater fuel filter OK?

YES NO

7

- Replace dirty flame heater fuel filter
  - Refer to Vol 2, chapter 4, para 4-72 or 4-79, and chapter 5, para 5-23

GO

TA 120838

Figure 8-1 (sheet 2 of 7)

8

- Put back flame heater fuel filter
  - Refer to Vol 2, chapter 4, para 4-72 or 4-79 and chapter 5, para 5-23

9

- Test flame heater nozzle
  - Unscrew two fuel line fittings. Refer to figure 9-1, 9-2, 9-3, or 9-4
  - Unscrew and shake out flame heater nozzle
  - Screw fuel line fittings into nozzle and tighten

- Put a bucket or can under flame heater nozzle
- Turn on heater for about 15 seconds, then turn heater off
- See if there is fuel in bucket or can

Is there fuel in bucket or can?

YES — GO

NO — **10**

- Replace flame heater nozzle
  - Refer to Vol 2, chapter 3, para 3-20 and chapter 5, para 5-23

GO

TA 120839

Figure 8-1 (Sheet 3 of 7)

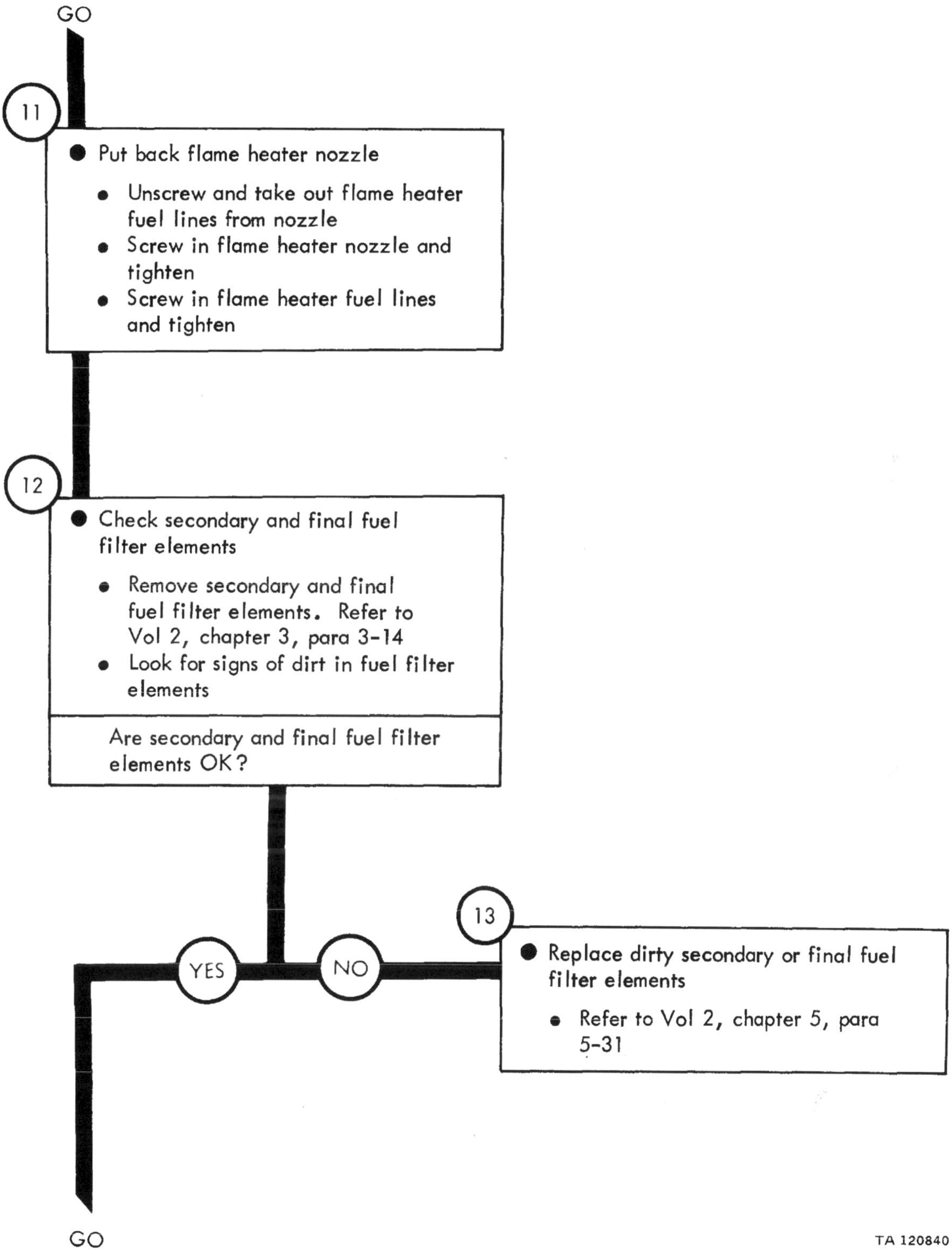

TA 120840

Figure 8-1 (Sheet 4 of 7)

GO

NOTE

The following procedure will need the use of two soldiers. The lead soldier will be called Soldier A. The helper will be called Soldier B

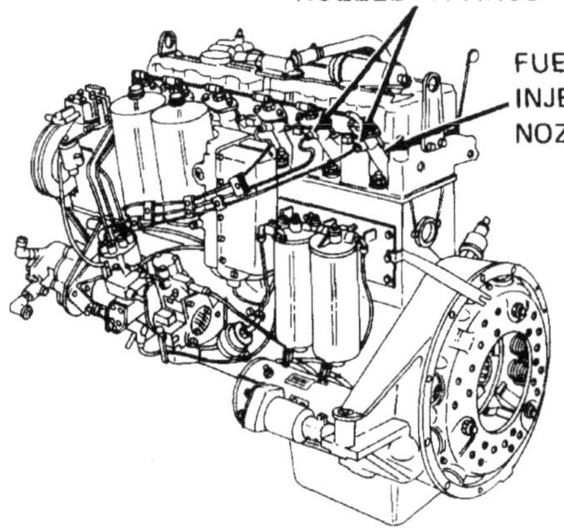

14

- Check for fuel supply from fuel injection pump
  - SOLDIER A: Loosen one fuel injector nozzle fitting but do not take it off
  - SOLDIER B: Crank engine for 5 seconds. Refer to Vol 2, chapter 6, para 6-6
  - SOLDIER A: See if fuel comes out of fuel injector nozzle fitting

Does fuel come out of fitting?

YES → 16 | NO → 15

15

- Replace fuel injection pump
  - Refer to Vol 2, chapter 3, para 3-10, and chapter 5, para 5-33

16

- Take out all six injector nozzles
  - Refer to Vol 2, chapter 3, para 3-18
- Clean, inspect and repair all six injector nozzles
  - Refer to Vol 2, chapter 4, para 4-40

Are injector nozzles OK?

GO

TA 120841

Figure 8-1 (Sheet 5 of 7)

GO

YES — NO

17

- Replace bad injector nozzles
  - Refer to Vol 2, chapter 5, para 5-34

18

- Put back all injector nozzles except the number 6 injector nozzle
  - Refer to Vol 2, chapter 5, para 5-34 and para 5-35

    Note: One injector nozzle will be needed for the next step

WARNING

Fuel coming out of an injector nozzle under pressure can go through the skin. This can cause blood poisoning. Keep hands away from injector nozzle when doing the next step

19

- Put number 6 fuel injector nozzle onto fuel supply tube
  - SOLDIER A: Connect nozzle to fuel supply tube. See figure 9-4
  - SOLDIER B: Crank engine for 5 seconds
  - SOLDIER A: See if correct fuel spray pattern comes from injector nozzle. Refer to figure 9-4

Is fuel spray pattern OK?

GO

TA 120842

Figure 8-1 (Sheet 6 of 7)

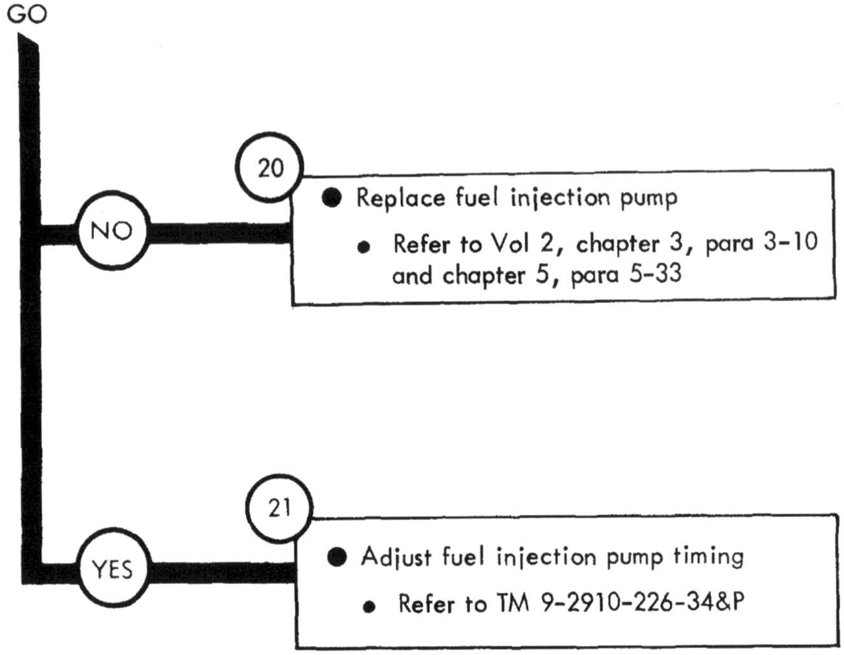

Figure 8-1 (Sheet 7 of 7)

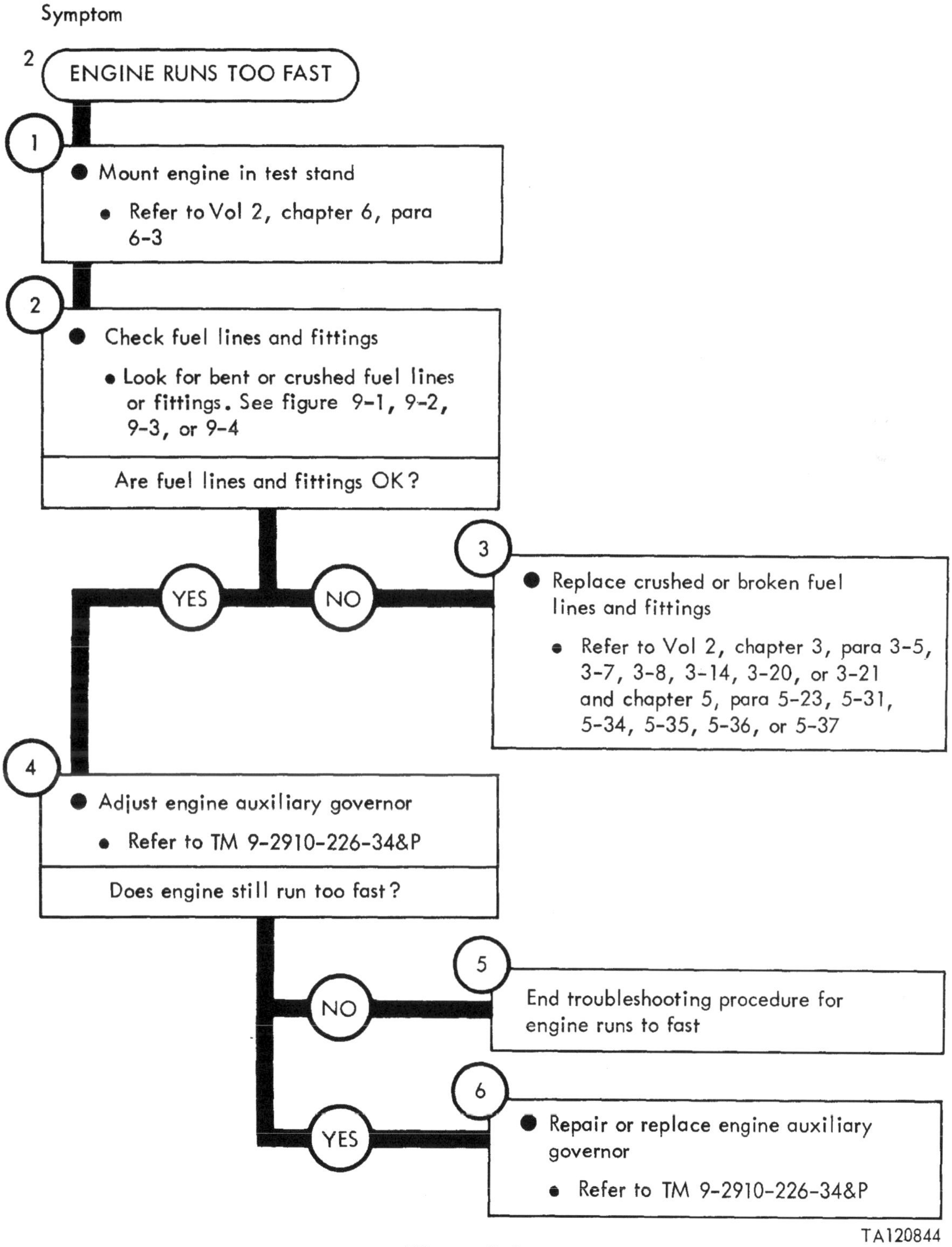
Symptom
2 ENGINE RUNS TOO FAST
1
Mount engine in test stand
Refer to Vol 2, chapter 6, para 6-3
2
Check fuel lines and fittings
Look for bent or crushed fuel lines or fittings. See figure 9-1, 9-2, 9-3, or 9-4
Are fuel lines and fittings OK?
YES
NO
3
Replace crushed or broken fuel lines and fittings
Refer to Vol 2, chapter 3, para 3-5, 3-7, 3-8, 3-14, 3-20, or 3-21 and chapter 5, para 5-23, 5-31, 5-34, 5-35, 5-36, or 5-37
4
Adjust engine auxiliary governor
Refer to TM 9-2910-226-34&P
Does engine still run too fast?
NO
5
End troubleshooting procedure for engine runs to fast
YES
6
Repair or replace engine auxiliary governor
Refer to TM 9-2910-226-34&P
TA120844

Figure 8-2

Symptom

3 WATER TEMPERATURE GAGE READS ABOVE 200°F WHILE RUNNING

1

- Mount engine in test stand
  - Refer to Vol 2, chapter 6, para 6-3

2

- Check water pump drive belts
  - Look for broken belts
- Check water pump drive belt adjustment
  - Refer to TM 9-2320-209-20 or TM 9-2320-211-20

Are water pump drive belt OK?

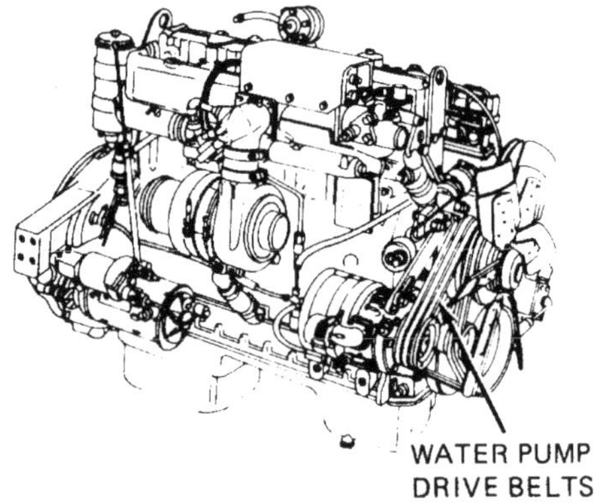

YES — GO

NO — **3**

- Replace broken water pump drive belt
  - Refer to TM 9-2320-209-20 or TM 9-2320-211-20
- Adjust loose drive belts
  - Refer to TM 9-2320-209-20 or TM 9-2320-211-20

GO

TA120845

Figure 8-3 (Sheet 1 of 4)

GO

4

- Check water pump
  - See if water pump leaks

Is water pump OK?

YES — go to 6

NO — go to 5

5

- Replace leaking water pump
  - Refer to Vol 2, chapter 3, para 3-27 and chapter 5, para 5-19
- Tighten clamps on leaking coolant hoses
  - Tighten clamp screw

    Note: If coolant does not stop leaking, the leaking hoses should be replaced. Refer to TM 9-2320-209-20 or TM 9-2320-211-20
- Replace bent or crushed coolant hoses
  - Refer to TM 9-2320-209-20 or TM 9-2320-211-20

6

- Check expansion plugs for water leaks
  - Look for signs of rust or water stains coming from expansion plugs
  - Look for cracks in engine block around expansion plugs

Are expansion plugs and engine block OK?

GO

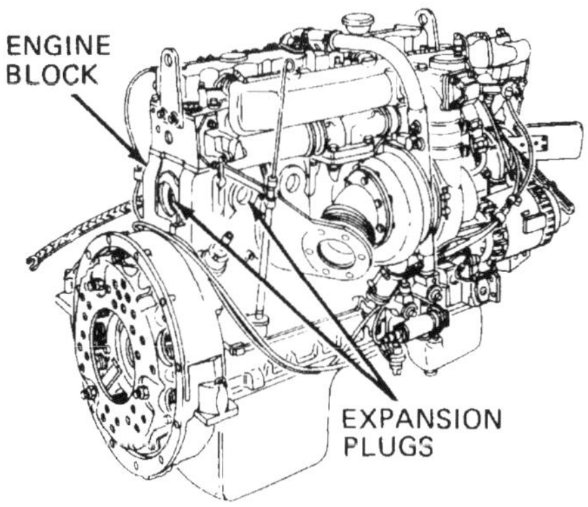

TA 120846

Figure 8-3 (Sheet 2 of 4)

GO

YES

NO

7

- Replace bad expansion plugs
  - Refer to Vol 2, chapter 4, para 4-6
- Replace cracked engine block
  - Refer to Vol 2, chapter 4, para 4-6

WARNING

Before going to the next step, make sure engine has had at least an hour to cool down before working on cooling system

8

- Test thermostat
  - Remove thermostat. Refer to Vol 2, chapter 3, para 3-27
  - Place thermostat in a can of water
  - Place thermometer in water
  - Heat water to 170° and see if thermostat starts to open
  - Heat water to 195° and see if thermostat fully opens

Is thermostat OK?

GO

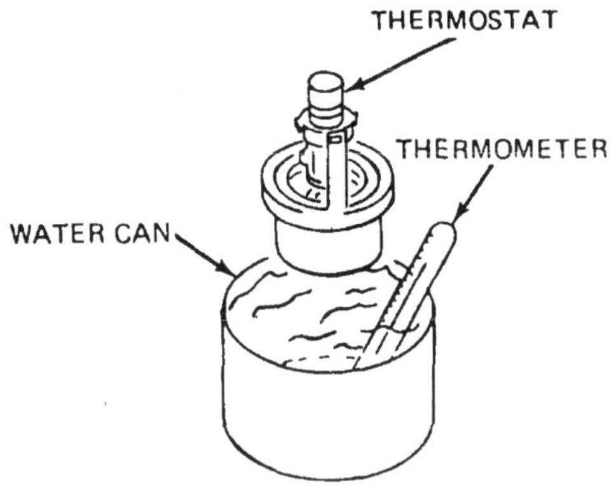

TA 120847

Figure 8-3 (Sheet 3 of 4)

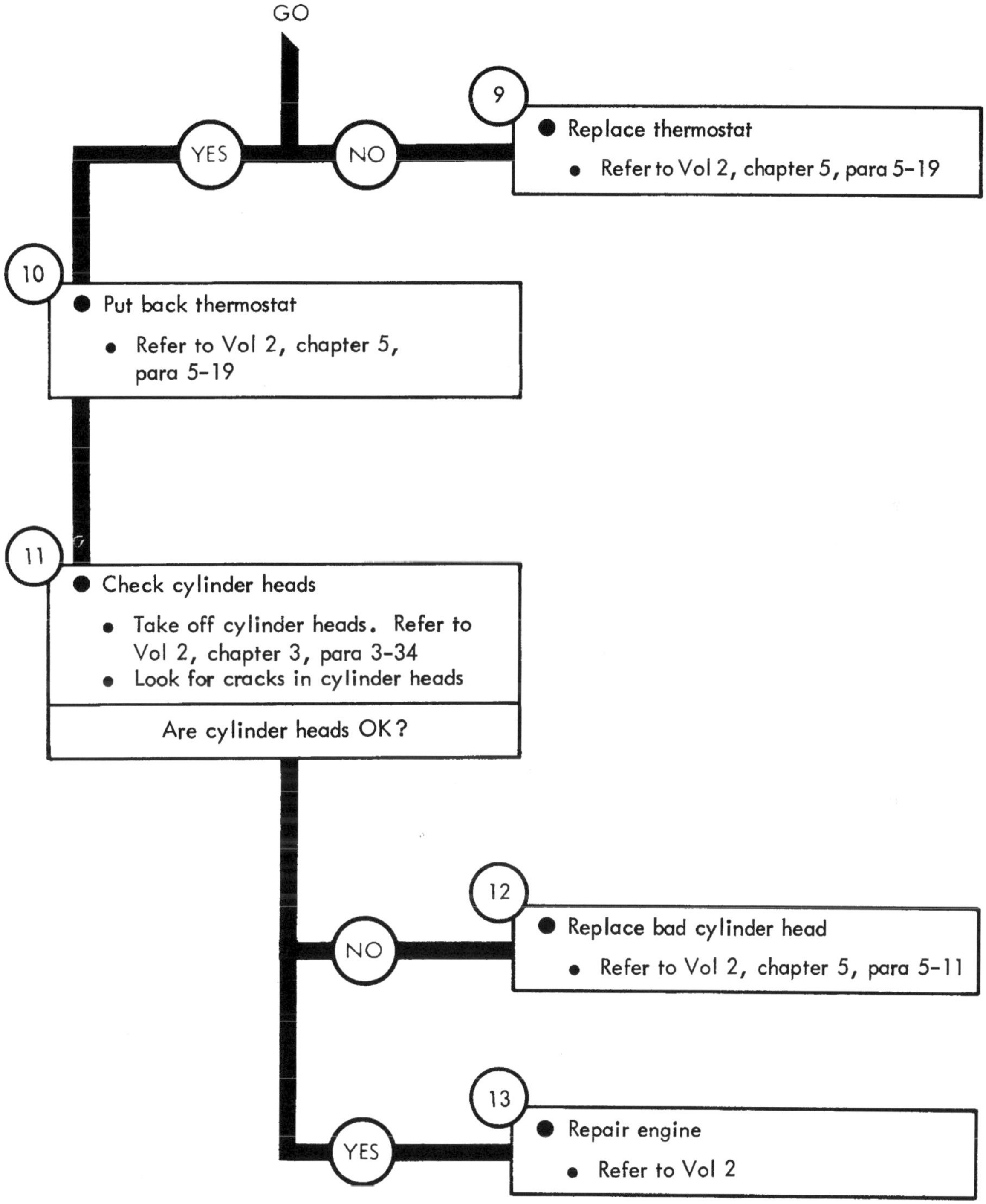

TA 120848

Figure 8-3 (Sheet 4 of 4)

Symptom

4 LOW OIL PRESSURE

1

- Mount engine in test stand
  - Refer to Vol 2, chapter 6, para 6-3

2

- Check oil filter gaskets
  - Look for signs of oil leaks at bottom of oil filter body
  - Feel for loose center bolt
- Check oil filter drain plug
  - Look for signs of oil leaking from drain plug

Are oil filter gaskets and oil filter drain plug OK?

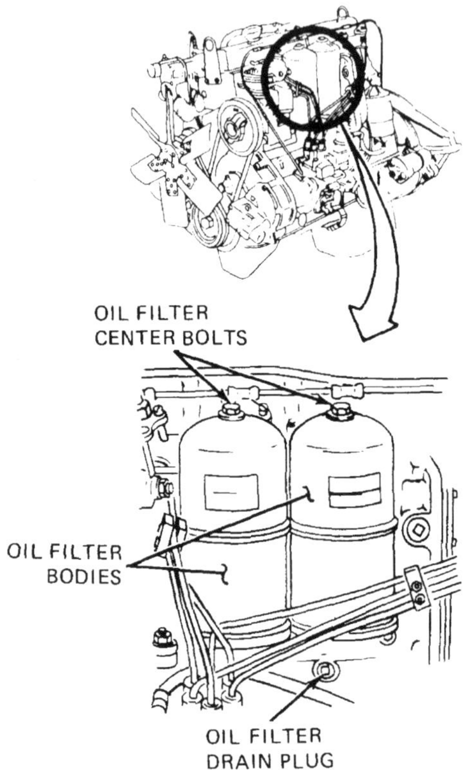

YES

NO

3

- Tighten loose oil filter center bolt
  - Tighten center bolt to 60 pound-feet
- Tighten loose oil filter drain plug
  - Refer to Vol 2, chapter 5, para 5-28

    Note: If oil should still leak at bottom of oil filter body replace oil filter gasket. Refer to Vol 2, chapter 3, para 3-11 and chapter 5, para 5-28

TA 120849

Figure 8-4 (Sheet 1 Of 5)

GO

4

- Check engine oil pan drain plug
  - Look for signs of oil leaking from drain plug

Is engine oil pan drain plug OK?

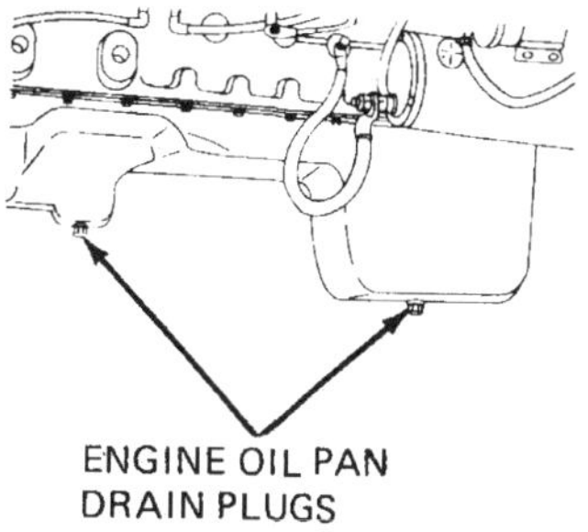

YES / NO

5

- Tighten loose engine oil pan drain plug
  - Tighten oil pan drain plug to 10 pound-feet

6

- Check for sticking oil pressure regulator valve
  - Take off oil pressure regulator valve housing. Refer to Vol 2, chapter 3, para 3-24
  - Take out pressure regulator valve and feel if plunger slides freely in bore. Refer to Vol 2, chapter 4, para 4-18

Does plunger slide freely?

YES / NO

7

- Repair scratches or burrs in oil pressure regulator valve housing
  - Refer to Vol 2, chapter 4, para 4-18
- Replace bad plunger
  - Refer to Vol 2, chapter 4, para 4-18

GO

TA 120850

Figure 8-4 (Sheet 2 of 5)

GO

8
- Put back oil pressure regulator valve housing
  - Refer to Vol 2, chapter 5, para 5-21

9
- Take off engine oil pan
  - Refer to Vol 2, chapter 3, para 3-39
  - See if pressure oil pump pickup tube screen is clogged

Is screen OK?

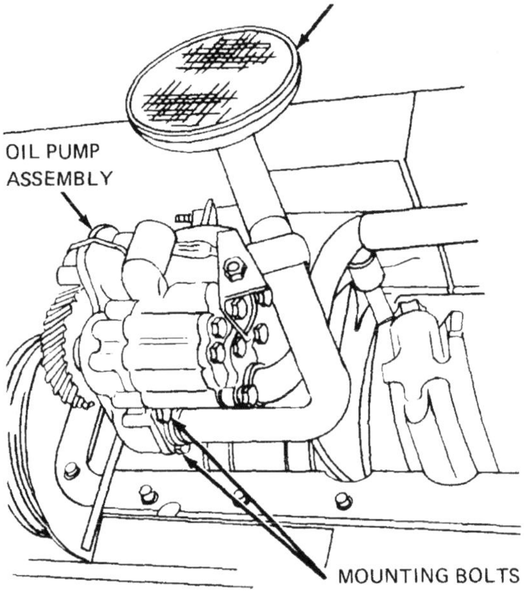

YES | NO

10
- Clean pressure oil pump pickup tube screen
  - Refer to Vol 2, chapter 4, para 4-17

11
- Check for loose pressure oil pump pickup tube mounting bolts

Are bolts tight?

GO

TA 120851

Figure 8-4 (Sheet 3 of 5)

GO

YES NO

12

● Tighten loose mounting bolts

13

● Check for loose pressure oil pump outlet tube mounting bolts at oil pump and engine block

Are mounting bolts tight?

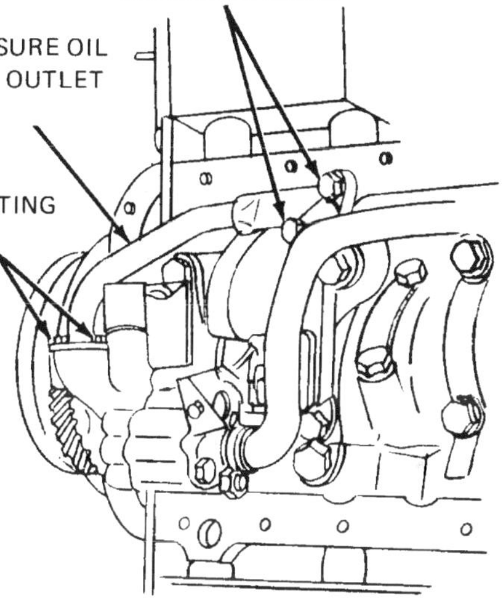

YES NO

14

● Tighten loose mounting bolts

GO

TA 120852

Figure 8-4 (Sheet 4 of 5)

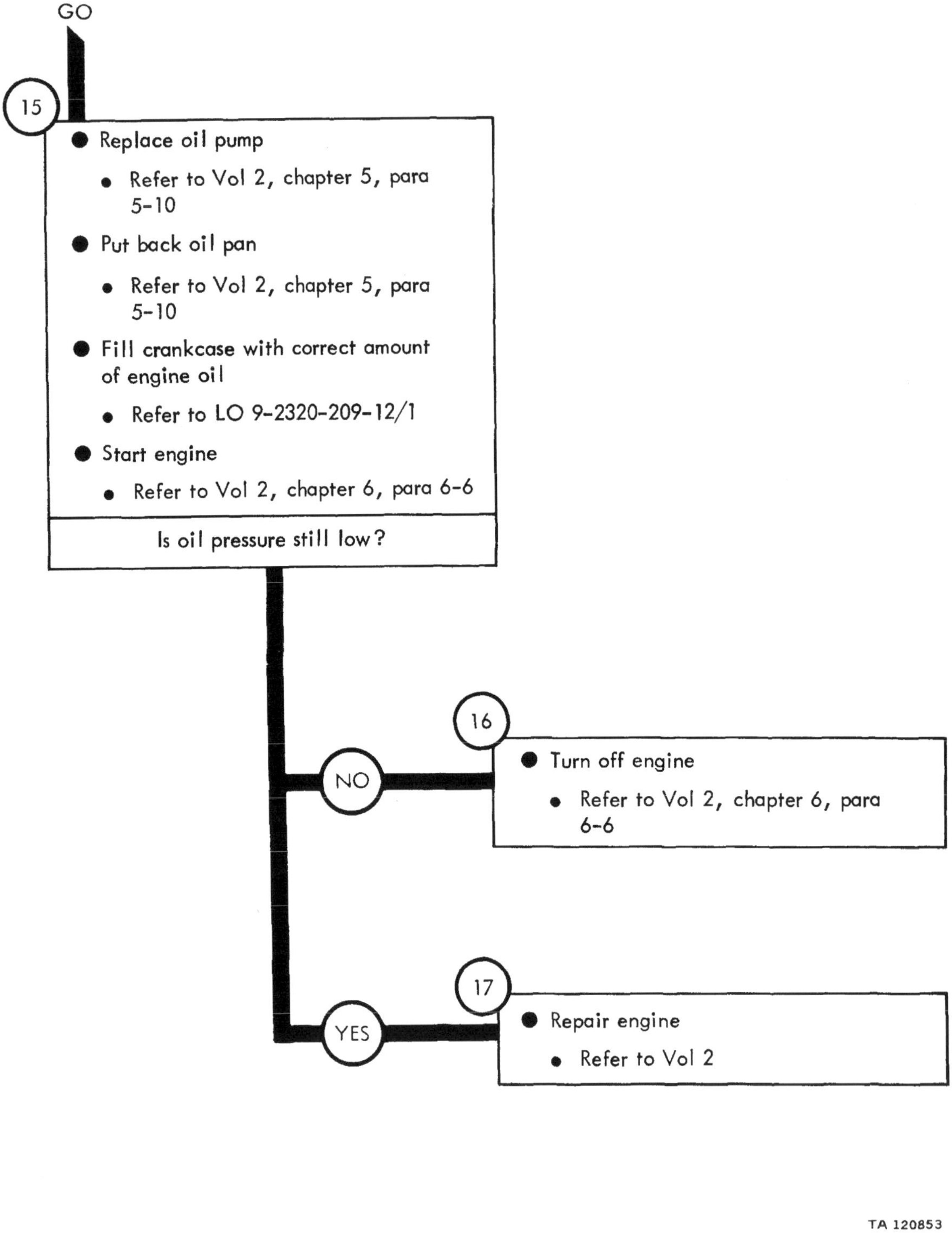

Figure 8-4 (Sheet 5 of 5)

Symptom

5 ENGINE LACKS POWER

WARNING

Diesel fuel is very flammable. Care must be used when choosing a place to work on fuel system. Keep engine about 50 feet away from an area where open flame, sparks, or smoking may cause a fire. Keep a fire extinguisher close by

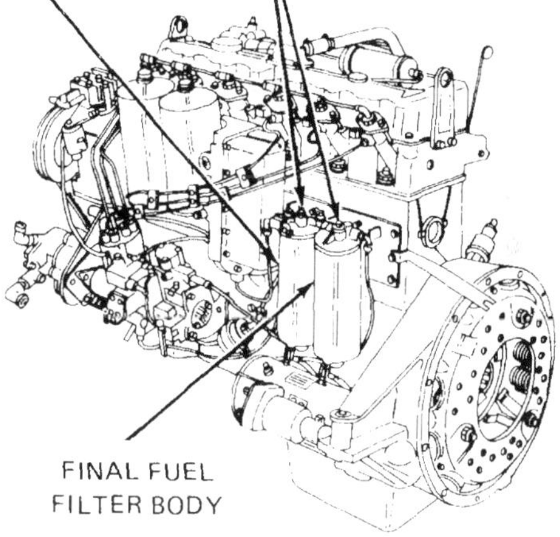

1

- Mount engine in test stand
  - Refer to Vol 2, chapter 6, para 6-3

2

- Check secondary and final fuel filter bodies
  - Look for fuel leaking from fuel filter bodies
  - Look for signs of fuel leaking from fuel filter head gasket
  - Feel for loose retaining screw

Are fuel filter bodies OK?

YES — GO

NO

3

- Tighten fuel filter bodies
  - Tighten retaining screws
- Replace leaking fuel filter head gaskets
  - Refer to Vol 2, chapter 3, para 3-14 and chapter 4, para 4-33 and chapter 5, para 5-31
- Replace cracked or broken fuel filter bodies
  - Refer to Vol 2, chapter 3, para 3-14, chapter 4, para 4-33, chapter 5, para 5-31

GO

TA 120854

Figure 8-5 (Sheet 1 of 3)

GO

4

- Check all fuel lines and fittings
  - Look for a crushed, or broken line. See figure 9-1, 9-2, 9-3 or 9-4
  - Look for signs of fuel leaking from lines or fittings
  - Feel for loose fittings

Are fuel lines and fittings OK?

YES — NO

5

- Tighten loose or leaking fuel line fittings
- Replace broken or crushed fuel lines
  - Refer to Vol 2, chapter 3, para 3-5, 3-7, 3-8, 3-18, 3-20, and 3-21, and chapter 5, para 5-23, 5-31, 5-33, 5-34, 5-35, 5-36, and 5-37

6

- Check secondary and final fuel filter elements
  - Remove secondary and final fuel filter elements. Refer to Vol 2, chapter 3, para 3-14 and chapter 4, para 4-33
  - Look for signs of dirt in fuel filter elements

Are secondary and final fuel filter elements OK?

YES — NO

7

- Replace dirty secondary or final fuel filter elements
  - Refer to Vol 2, chapter 4, para 4-33 and chapter 5, para 5-31

GO

TA120855

Figure 8-5 (Sheet 2 of 3)

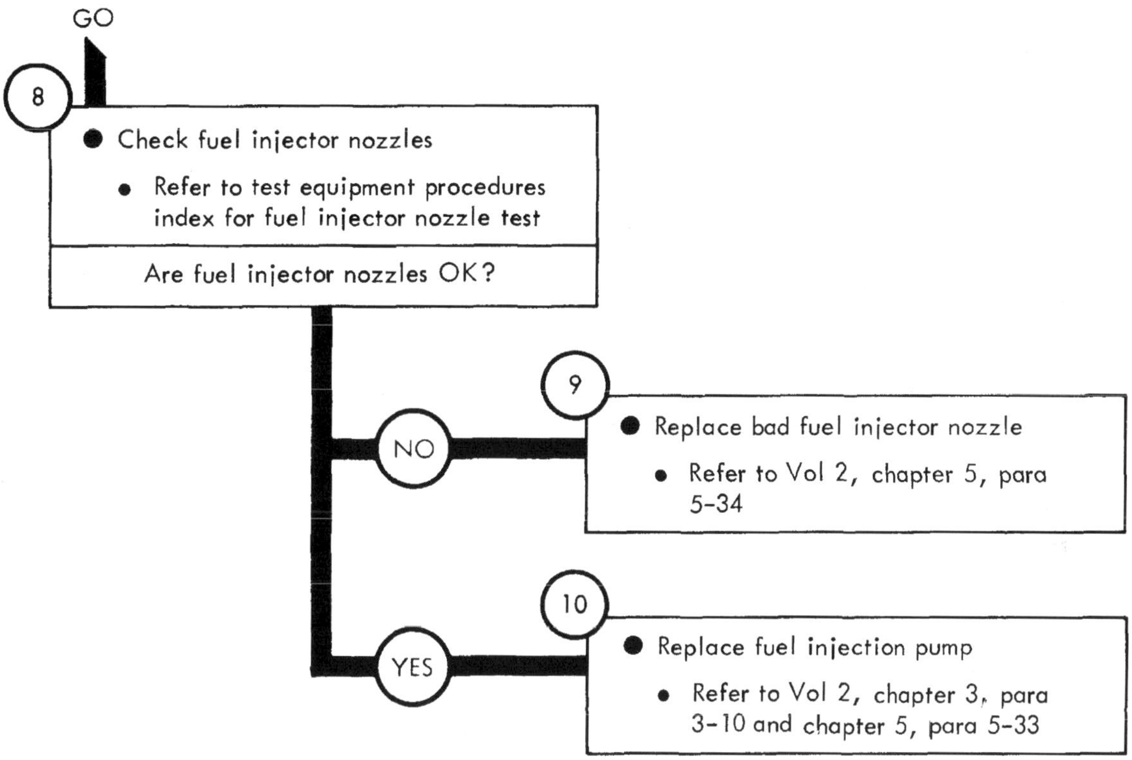

Figure 8-5 (Sheet 3 of 3)

CHAPTER 9

ENGINE SYSTEM SUPPORT DIAGRAMS

9-1.GENERAL. This chapter gives the diagrams you need when doing trouble-shooting procedures in chapter 8. Figure 3-1 is a complete listing of all support diagrams used in this manual.

154-882 0-94-4

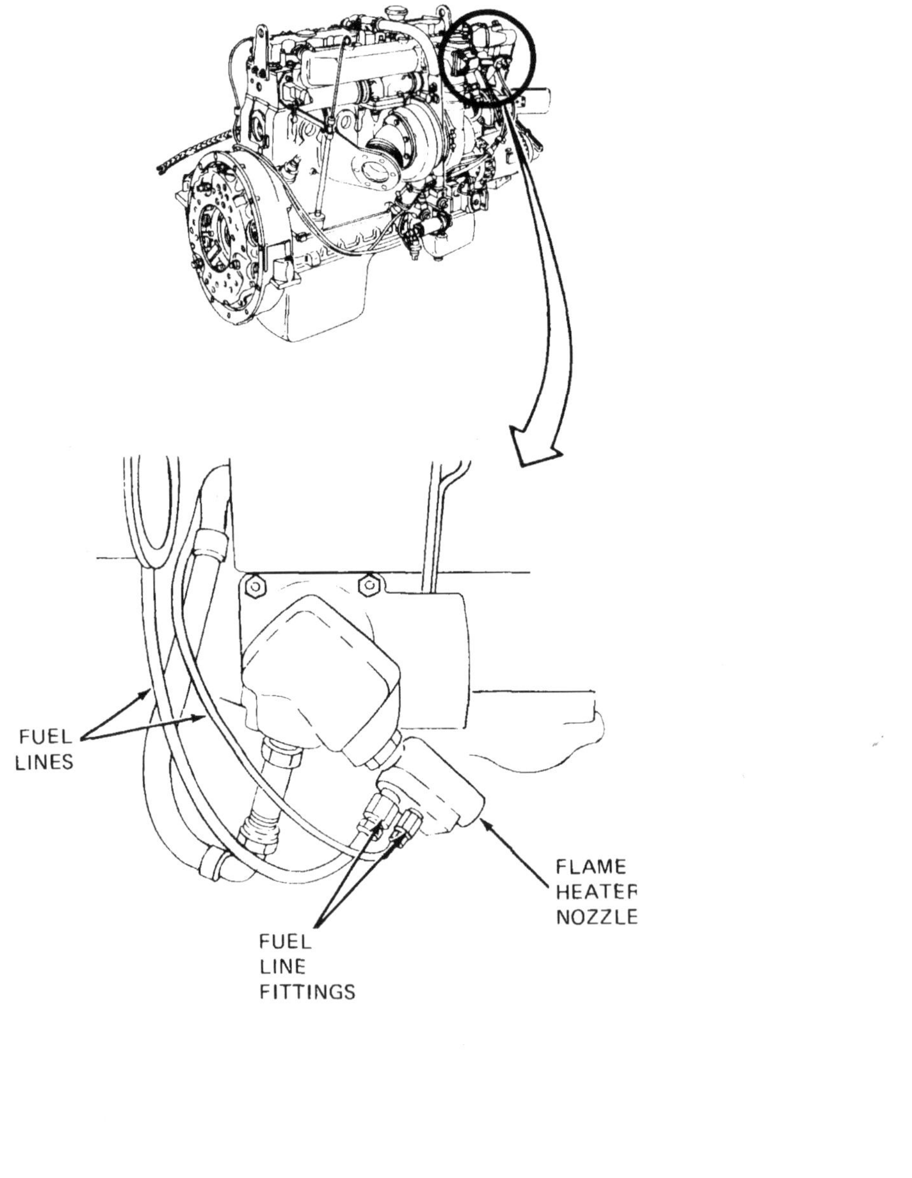

Figure 9-1. Flame Heater Nozzzle, Fuel Lines, and Fittings
for LD-465-1 and LD-465-1C

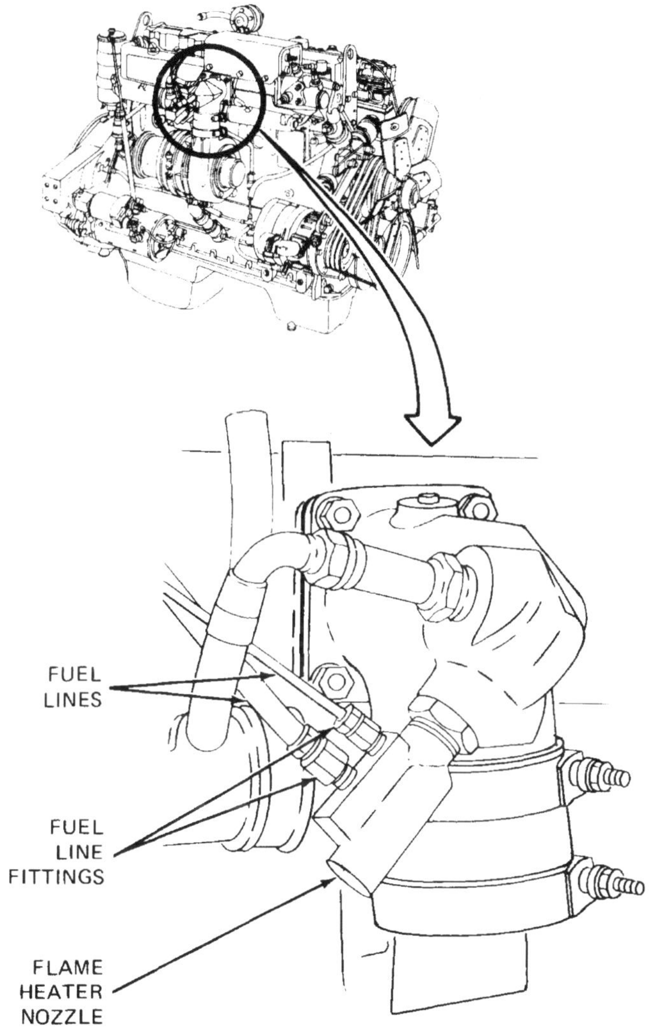

Figure 9-2. Flame Heater Nozzle, Fuel Lines, and Fittings for LDS-465-1, LDS-465-1A and LDS-465-2

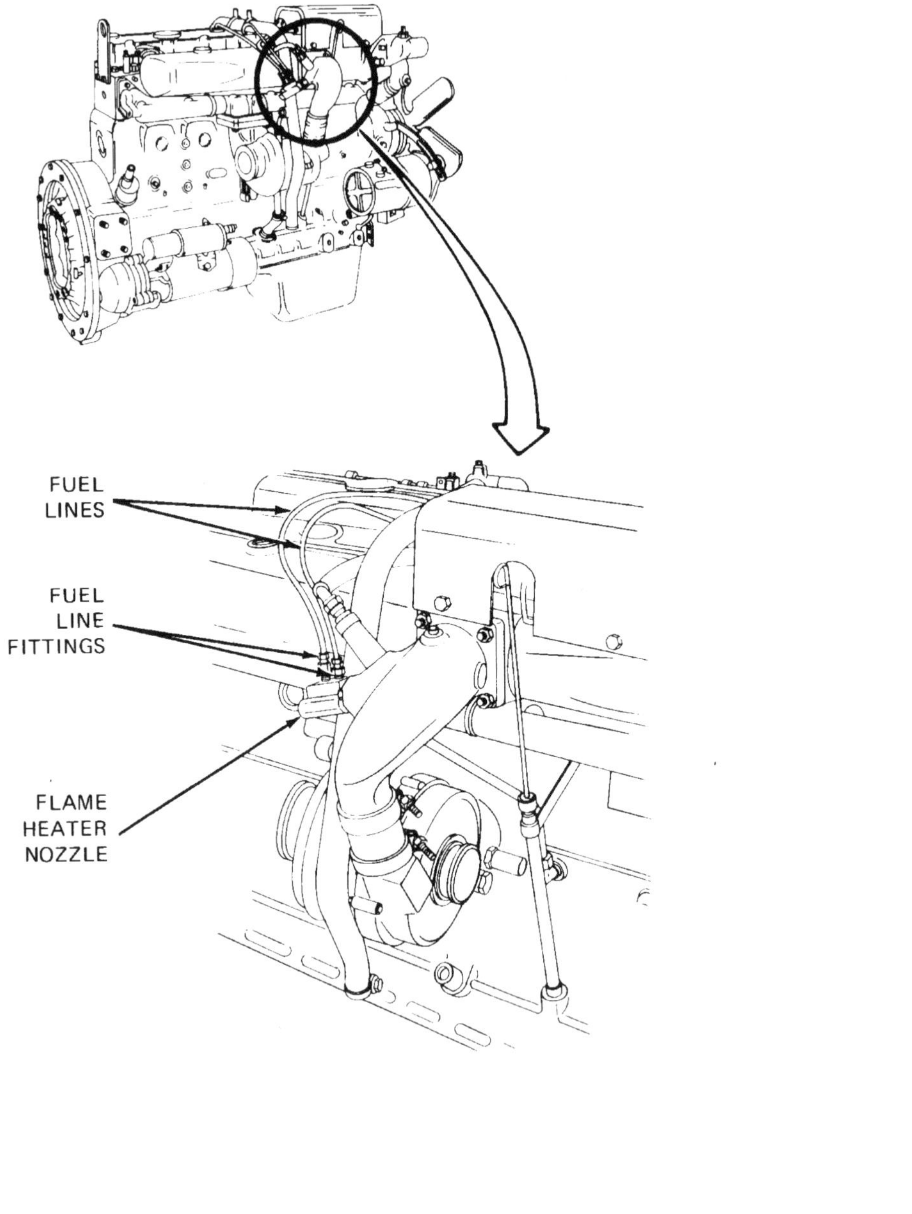

FIGURE 9-3. Flame Heater Nozzle, Fuel Lines, and Fittings for LDT-465-1

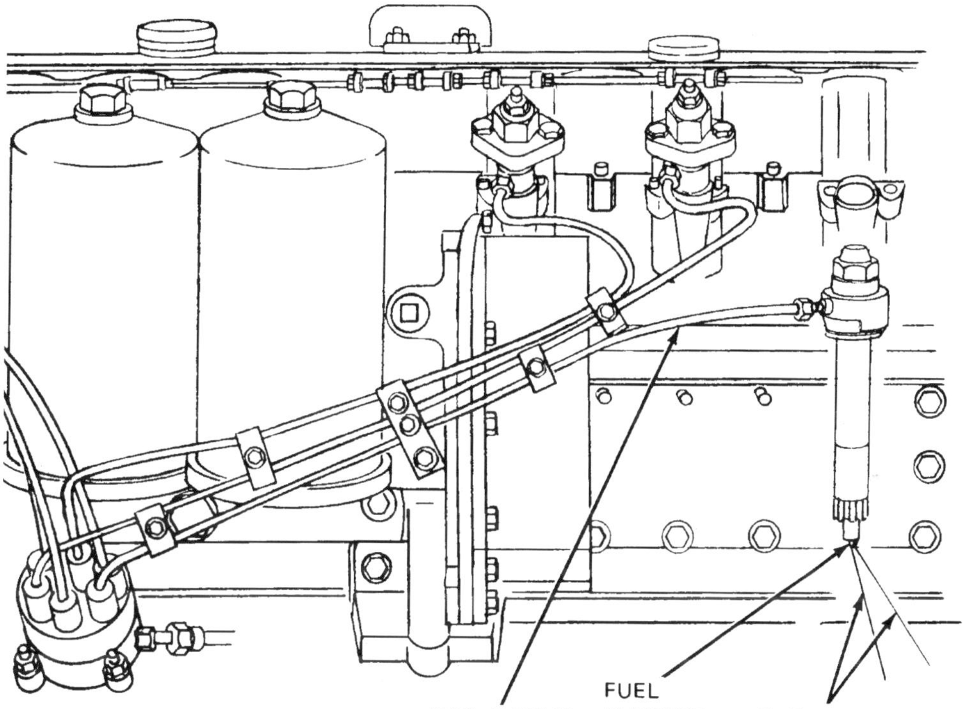

Figure 9-4. Fuel Injector Nozzle Spray Pattern

154-882 0-94-5

CHAPTER 10

ENGINE SYSTEM TEST PROCEDURE

10-1. GENERAL. This chapter gives test procedure for the tests given in chapter 8, for the engine system.

10-2. TEST SET-UP. Instructions for setup of test equipment and parts to be tested are given before the test procedures. Illustrations are used, when needed, to show you how to hook up the test equipment to the part to be tested.

10-3. TEST PROCEDURE. Detailed step-by-step instructions, in flow chart form, are given for each test. The procedure calls out the type of test and the condition of the engine system for each part of testing. The step-by-step test will lead you to the bad component or to a fault symptom within a related system. Reference is made to the fault symptom index, chapter 6, if the test shows a fault in another system.

FUEL INJECTOR NOZZLE TEST

1

- Start and warm up engine
  - Refer to Vol 2, chapter 6, para 6-6

WARNING

Diesel fuel is very flammable. Care must be used when choosing a place to work on fuel system. Keep engine about 50 feet away from an area where open flame, sparks, or smoking can cause a fire. Keep a fire extinguisher close by

NOTE

Put a rag under line fittings before loosening to catch any fuel which drips out

GO

TA 120860

Figure 10-1 (Sheet 1 of 2)

GO

2

- Check fuel injector nozzles
  - Loosen fuel injector nozzle fitting of one fuel injector nozzle and see if engine runs rougher

    Note: If engine runs rougher the injector nozzle is OK
  - Tighten fuel injector nozzle fitting

    Note: Do procedure for all six fuel injector nozzle fittings
- Stop engine
  - Refer to Vol 2, chapter 6, para 6-6

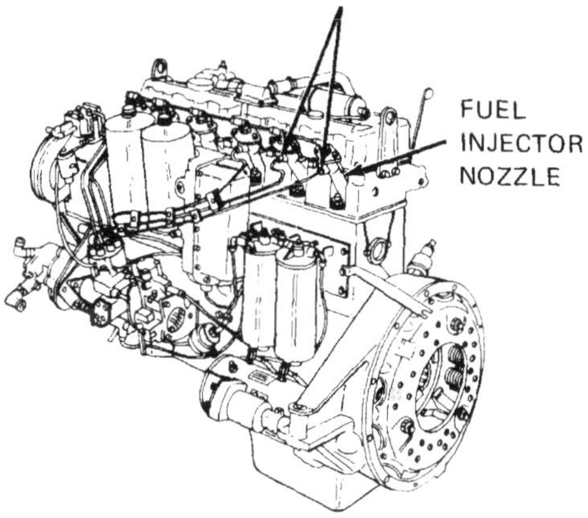

TA 120861

Figure 10-1 (Sheet 2 of 2)

By Order of the Secretaries of the Army and the Air Force:

E. C. MEYER
General, United States Army
Chief of Staff

Official:

J. C. PENNINGTON
Major General, United States Army
The Adjutant General

LEW ALLEN, JR., General, USAF
Chief of Staff

Official:

VAN L. CRAWFORD, JR., Colonel, USAF
Director of Administration

Distribution:

To be distributed in accordance with DA Fom 12-38, Direct and General Support Maintenance requirements for 2-1/2-Ton Truck Cargo, 2-1/l-Ton Truck Van, 5-Ton Truck Chassis, 5-Ton Truck Van, and 5-Ton Truck Cargo, etc.

U.S. GOVRNMENT PRINTING OFFICE : 1994 0-154-882

THE METRIC SYSTEM AND EQUIVALENTS

LINEAR MEASURE

1 Centimeter = 10 Millimeters = 0.01 Meters = 0.3937 Inches
1 Meter = 100 Centimeters = 1000 Millimeters = 39.37 Inches
1 Kilometer = 1000 Meters = 0.621 Miles

WEIGHTS

1 Gram = 0.001 Kilograms = 1000 Milligrams = 0.035 Ounces
1 Kilogram = 1000 Grams = 2.2 Lb
1 Metric Ton = 1000 Kilograms = 1 Megagram = 1.1 Short Tons

LIQUID MEASURE

1 Milliliter = 0.001 Liters = 0.0338 Fluid Ounces
1 Liter = 1000 Milliliters = 33.82 Fluid Ounces

SQUARE MEASURE

1 Sq Centimeter = 100 Sq Millimeters = 0.155 Sq Inches
1 Sq Meter = 10,000 Sq Centimeters = 10.76 Sq Feet
1 Sq Kilometer = 1,000,000 Sq Meters = 0.386 Sq Miles

CUBIC MEASURE

1 Cu Centimeter = 1000 Cu Millimeters = 0.06 Cu Inches
1 Cu Meter = 1,000,000 Cu Centimeters = 35.31 Cu Feet

TEMPERATURE

$5/9\,(^{0}F - 32) = {}^{0}C$
212^{0} Fahrenheit is equivalent to 100^{0} Celsius
90^{0} Fahrenheit is equivalent to 32.2^{0} Celsius
32^{0} Fahrenheit is equivalent to 0^{0} Celsius
$9/5\,C^{0} + 32 = F^{0}$

APPROXIMATE CONVERSION FACTORS

TO CHANGE	TO	MULTIPLY BY
Inches	Centimeters	2.540
Feet	Meters	0.305
Yards	Meters	0.914
Miles	Kilometers	1.609
Square Inches	Square Centimeters	6.451
Square Feet	Square Meters	0.093
Square Yards	Square Meters	0.836
Square Miles	Square Kilometers	2.590
Acres	Square Hectometers	0.405
Cubic Feet	Cubic Meters	0.028
Cubic Yards	Cubic Meters	0.765
Fluid Ounces	Milliliters	29.573
Pints	Liters	0.473
Quarts	Liters	0.946
Gallons	Liters	3.785
Ounces	Grams	28.349
Pounds	Kilograms	0.454
Short Tons	Metric Tons	0.907
Pound-Feet	Newton-Meters	1.356
Pounds per Square Inch	Kilopascals	6.895
Miles per Gallon	Kilometers per Liter	0.425
Miles per Hour	Kilometers per Hour	1.609

TO CHANGE	TO	MULTIPLY BY
Centimeters	Inches	0.394
Meters	Feet	3.280
Meters	Yards	1.094
Kilometers	Miles	0.621
Square Centimeters	Square Inches	0.155
Square Meters	Square Feet	10.764
Square Meters	Square Yards	1.196
Square Kilometers	Square Miles	0.386
Square Hectometers	Acres	2.471
Cubic Meters	Cubic Feet	35.315
Cubic Meters	Cubic Yards	1.308
Milliliters	Fluid Ounces	0.034
Liters	Pints	2.113
Liters	Quarts	1.057
Liters	Gallons	0.264
Grams	Ounces	0.035
Kilograms	Pounds	2.205
Metric Tons	Short Tons	1.102
Newton-Meters	Pound-Feet	0.738
Kilopascals	Pounds per Square Inch	0.145
Kilometers per Liter	Miles per Gallon	2.354
Kilometers per Hour	**Miles per Hour**	**0.621**

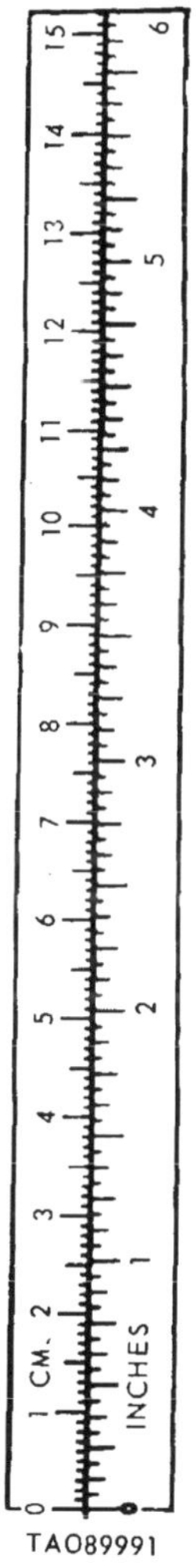

TA089991

048856

Made in United States
Orlando, FL
04 November 2026

90202941R00035